THE PRENTICE HALL
IT CAREER GUIDE

THE PRENTICE HALL IT CAREER GUIDE

Fran Quittel

Upper Saddle River, New Jersey 07458

Executive editor: Bob Horan
Project Manager: Lori Cerreto
Production editor: Wanda Rockwell
Manufacturer: Von Hoffmann Graphics

ISBN 0-13-101933-3

10 9 8 7 6 5 4 3 2 1

TABLE OF CONTENTS

ABOUT THE AUTHOR

Fran Quittel created and ran the first career forum on the Microsoft Network (MSN) and is known on the web as "CareerBabe" via the www.careerbabe.com, a site which provides career advice to students and adults alike. She is also the author of *FirePower! Everything You Need to Know Before and After You Lose Your Job* (10 Speed Press, 1994), and has been Computerworld's Career Adviser for the past three years.

Yale-educated with lifetime teaching credentials, Fran is bilingual (Spanish-English) and a former university linguist. She has spent the past 20 years recruiting for technology companies across engineering, software development, sales, marketing and executive positions. Additionally, she is viewed as an expert on the systems side of recruiting, re-engineering the function to reduce expenses dramatically, while working across multiple industries, including technology, education and healthcare. With lifetime teaching credentials, Fran has also been involved in education, upgrading teacher recruitment and working on school to career programs for both adults and younger students.

Fran serves on the Board of the Open Door Foundation, the non-profit association of the National Association of Computer Consulting Businesses, which provides scholarships to students of any age to encourage continuing studies in computer science. For fun, she writes, hikes, practices yoga, and is learning to fly fish.

f

1. INTRODUCTION

Today's high-tech job market has taken a radical turn from the boom that overtook the industry from 1994 – 2000. While just 18 months ago, college graduates with computer science and business degrees frequently left school with multiple offers after short job searches, today's student must spend months prior to graduation developing leads, networking, writing resumes, and following up. They must spend summers prior to graduation working on internships with companies that may offer permanent employment. In short, there's not a moment to waste.

Nonetheless, even this tough job market has its benefits. An invaluable lesson to learn early in your career is this: most problems do have a solution and sometimes thinking outside the box pulls off miracles. The information provided in this introduction to the job-search process should help you conquer the job search blues. It provides insights and creative strategies to help when you think you're stuck.

While economic shifts, downsizing, and events of pure chance prevent us from predicting exactly when you will find your first job, there is one certainty:

Sustained levels of certain kinds of activities do produce predictable outcomes. Therefore, continuing to look – even in the face of disappointment - increases the likelihood that you will find worthwhile work at a competitive salary.

In other words, while we cannot predict exactly *when* you will get a job, we do know that with a certain level of consistent effort – and some luck – you will find your first real post-college job.

The harder I work, the luckier I get.
Source: Samuel Goldwyn

My mission is to motivate you to overcome the tight job market blues -- to keep you returning to your job search. Never giving up is what will make your search pay off.

The material in this book will help you:

1. Make yourself into a skilled candidate in demand
2. Understand the power of networking so you have plenty of job leads from people with whom you have a relationship
3. Build a great resume tailored to each type of job opportunity you seek
4. Consider jobs in industries that need your skills
5. Be persistent

This book will give you what you need to get your search started on the right track. Let's start now.

2. MARKET OVERVIEW

While there is an abundance of resources online and off, in books, journals, and newspapers, there are specific sources to orient you to jobs across IT and business functions from programming to marketing. These sources include: *Tomorrow's Jobs*, published by the US Department of Labor's Bureau of Labor Statistics, describing growth and attrition in the total job market from year 2000 to 2010; and J. Michael Farr and LaVerne L. Ludden's *Best Jobs for the 20th Century*.[1]

Additionally, in May of 2002, the Information Technology Association of America published a new study called "Bouncing Back: Jobs, Skills and the Continuing Demand for IT Workers." Plus, the Occupational Outlook Handbook, 2002-2003 edition, while covering all professional and related occupations (http://www.bls.gov/oco/oco1002.htm), also contains a section covering computer and mathematical occupations http://www.bls.gov/oco/cg/cgs033.htm. All of these sources, plus material from industry association magazines, conferences, and Web sites -- not to mention information from NACE, the National Association of Colleges and Employers -- will orient you to the people employed in IT, their starting salaries, and job outlook. (See Exhibits 2.1 and 2.2.)

---

[1] "Best Jobs for the 21st Century," J. Michael Farr and LaVerne L. Ludden, Ed.D., copyright 2001. Published by JISTWorks, an imprint of Jist Publishing, Inc., Indianapolis, Indiana.

Exhibit 2.1
STARTING SALARIES FOR GRADS WITH BACHELOR'S DEGREES
SUMMARY DATA 2001
OCCUPATIONAL OUTLOOK HANDBOOK
2002-2003 EDITION

Data Source	JOB TITLE		SALARY
NACE[2]	Computer programmer – Bachelor's degree with a degree in computer programming		$48,602
RHI[3]	Applications development programmer/developer		$58,500 - $90,000
RHI	Software development programmer/analyst		$54,000 - $77,750
RHI	Internet programmer/analyst		$56,500 - $84,800
NACE	Bachelor's degree in computer engineering		$53,9924
NACE	Master's degree in computer engineering		$58.026
NACE	Bachelor's degree in computer science		$52,723
RHI	Software engineer working in software development		$62,750 - $92,000 plus profit sharing, stock, company car/mileage
RHI	Computer support: help desk		$30,500 - $56,000
RHI	Senior tech support		$48,000 - $61,000
RHI	Systems administration		$50,250 - $70,250
NACE	B.S. in computer science		$52,723
NACE	Masters in computer science		$61,453

(continued)

---

[2] NACE, National Associating of Colleges and Employers
[3] Robert Half International.

(continued)

NACE	B.S. in computer programming		$48,602
NACE	B.S. in computer systems analysis		$45,643
NACE	B.S. in information science and systems		$45,182
NACE	B.S. in management information systems		$45,585
RHI	Database administrators		$72,500 - $105,750
RHI	Webmasters		$58,000 - $82,500
RHI	Internet/intranet developers		$56,250 - $76,750
NACE	B.S., computer hardware engineer		$53,924
NACE	M.S., Computer hardware engineer		$58,026
NACE	Ph.D., computer hardware engineer		$70,140

Exhibit 2.2
PROFESSIONAL OUTLOOK[4]

# of people in this job in the year 2000	Job Title	Outlook
585,000	Programmer, includes 22,000 self-employed	As fast as average but slower growth than that of other computer specialists
697,000	Computer software engineers applications & systems. Includes: 380,000 in applications 317,000 in systems	Very fast growing
734,000	Includes: -506,000 computer support specialists (tech support/help desk) -229,000 network and systems administrators (network: LAN/WAN, Internet/Intranet	Very fast growing

(continued)

[4] *Source:* Occupational Outlook Handbook. US Department of Labor, Bureau of Labor Statistics, http://www.bls.gov/oco/oco1002.htm

(continued)

47,000	Operations research analysts	Slower than average growth; many government jobs here. Job title not continuing.
887,000	Systems analysts: 431,000 Computer scientists: 21,000 Database administrators: 106,000 Network systems and data communications analysts: 119,000 Other specialists: 203,000 Includes 71,000 self-employed	Very fast growing, faster than average
60,000	Computer hardware engineers	Fast growing

3. OVERALL JOB MARKET DATA

According to the Bureau of Labor Statistics, the total U.S. job market by the year 2010 will include approximately 168 million people, up from 146 million in the year 2000. Moreover, women's participation in the workforce will grow to 47.9%, up from 46.6% at the start of the decade, while the percentage of U.S. men in the workforce will drop from 53.4% at the start of the decade to 52.1% by its end. By 2010, young people age 16 to 24 will total 16.5% of the workforce, while people age 25 to 54 will drop from 71% of the workforce in year 2000 to 66.6% in 2010. Finally, although many baby boomers are slated to retire, more people over age 55 will continue working, raising this group's representation in the labor pool from 12.9% to 16.9%.

Not only will education be important for most people seeking work, in fact, says the Bureau of Labor Statistics, 3 out of every 5 jobs, or 13.7 million jobs, will be added within the service-producing sector of our economy. More than two-thirds of this job growth will be in business, health, and social services. Although social services will add 1.2 million jobs, and health services will add 2.8 million, growth in business services, including personnel supply services and computer and data processing services, will outstrip all other areas, adding 5.1 million jobs.

In other words, states the BLS, even if you are currently facing a tougher than usual search for your first position, "employment in computer and data processing services – which provides prepackaged and specialized software, data and computer systems design and management, and computer-related consulting services – is projected to grow by 86% between 2000 and 2010, ranking as the fastest growing industry in the economy."[5] Not only will computer-related jobs account for eight out of the fastest growing twenty occupations in the US economy, throughout its report, BLS predictions state that "nearly three-quarters of the job growth will come from three groups of professional occupations: computer and mathematical occupations; healthcare practitioners and technical occupations;

---

[5] Tomorrow's Jobs, Occupational Outlook Handbook 2002-03 Edition, US Department of Labor, Bureau of Labor Statistics, Feb. 2002, p.4.

and education, training, and library occupations." These groups
will add 5.2 million jobs combined. [6] (See Exhibit 3.1.)

Exhibit 3.1
PERCENT CHANGE IN EMPLOYMENT IN OCCUPATIONS
PROJECTED TO GROW FASTEST, 2000-10

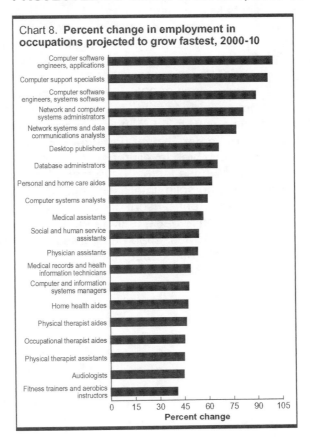

Chart 8. **Percent change in employment in occupations projected to grow fastest, 2000-10**

Source: BLS, p. 6.

---

[6] Ibid, p. 5.

4. SO WHAT ABOUT NOW?

Unfortunately, the period of 2000 – present has shown a nasty downturn in hiring overall. The dot-com bust of 2000, the recession, the end of Y2K projects, the events of September 11, and the ongoing revelations of corporate shenanigans and financial misconduct brought corporate spending for new technology projects to a screeching halt. In the boom years of 1995 – 2000, new technology grads often named their salary and sign-on bonus. Today, unemployment has skyrocketed, particularly in major metropolitan areas and in the West, the area of greatest upturn during dot-com and Y2K boom days. "In 2001," said *Infoweek* in January of 2002, "monthly IT unemployment rates ranged from 1.9% in April to 5% from August through October and then upward to 5.5% in November before settling back to 4.4% in December. Never had unemployment among IT workers stayed at such high levels for such an extended period." [7]

In short, all of these factors have made the year 2002 fairly tough in terms of finding solid, interesting opportunities. Students today need to be persistent. "I sent out many resumes and often didn't even get a response back," said Leeann Sobehart, University of Dayton business marketing graduate who did get her first job at the University of Pittsburgh.

---

[7] Information Week, 1-11-2002, "2001 IT Jobless Rate Was Highest Ever," by Eric Chabrow.

5. THE IT OUTLOOK

Within this overall turmoil, and despite its rosy long-term outlook, IT shed approximately 5% of its jobs overall from 2001 to 2002. A close examination of the data shows the recession hitting IT workers within technology companies more than three times as hard in terms of layoffs as compared with IT workers employed in non-IT companies. Exhibit 5.1 explains this trend.

Exhibit 5.1

Burst Bubble: The IT Workforce in 2001 and the Demand for IT Workers[8]

In response to economic conditions in the IT industry and elsewhere, the IT job market took a tumble in 2001. The 10.4-million-member IT workforce measured by ITAA at the beginning of 2001 fell to 9.9 million workers at the start of 2002. In 2001, the IT workforce had 9.5 million workers in non-IT companies, and 960,000 in IT companies. As of May 2002, those figures dropped to 9.1 million non-IT company workers and 800,000 IT company workers. In percentage terms, as of May 2002, the year was far harder for those within IT companies. IT firms lost 15 percent of their IT workers, compared to four percent for non-IT firms.

Then there is the question of salaries and bonuses. "Information technology managers and IT staff saw their total compensation decline in 2002 by 8% and 11% respectively," reported *InformationWeek*.[9] A May 2002 survey from Challenger, Gray and Christmas reported a "dramatic" decline in bonuses for IT staff members, from a median of $11,000 in 2001 to a mere $2,000 in 2002.[10]

Right along with reductions in job opportunities for their more experienced colleagues, internships for technology majors

---

[8] Bouncing Back, p. 11.
[9] Information Week, 2002 Compensation Survey. "2002 Tech Jobs News Good, Bad."
[10] John A. Challenger, Challenger, Gray & Christmas, inc., 5-10-2002.

and new jobs for seniors dropped significantly along with the economy overall.[11] Except for the Federal Government and education, two of the few industry segments predicting an upswing in new grad hiring, NACE's *Job Outlook 2002* held some disturbing news: For 2001 -2002 grads, NACE predicted a 36.4% drop in hiring. However, for 2003, NACE predicts only a 3.6% decline in hiring, down from the 2001 - 2002 figures.

The Good News

But close scrutiny of both today's outlook as well as of the long-term does provide a positive picture for technology graduates. First, getting a bachelor's, master's or Ph.D. in technology, computer science, computer engineering, IT, MIS, programming, or business, with a computer concentration, good grades, and experience through an internship will put you in the catbird seat. Overall, new graduates with majors in computer science, computer engineering, information sciences and systems, business administration, and management information systems are still the most desired new hires in all areas of the country. IT companies and non-IT companies, such as those in healthcare and the insurance industry, are thinking about filling some 1 million jobs in addition to some 580,000 IT jobs that go unfilled due to a "gap" of qualified talent.

Second, computer science, computer engineering, and technical graduates remain at the top of the wage-earner pyramid, outearning their liberal arts siblings by as much as $25,000 - $30,000 in their first jobs. This means that despite the drop in *total* compensation, losing such perks as stock options, generous sign-on bonuses, and even new cars – all common during dot-com boom years -- base salaries among this group of graduates remain substantially above what most other new graduates earn.

In fact, in its Summer 2002 study of new graduate starting salaries, the National Association of Colleges and Employers came to these conclusions: 2002 liberal arts graduates earned an average starting salary of under $30,000 but graduates in fields ranging from MIS to computer engineering earned average

---

[11] Bouncing Back: Jobs, Skills and the Continuing Demand for IT Workers, ITAA, including the New ITAA/DICE Tech Skills Profile, May 2002.

starting compensations ranging from $42,705 to $51,387 respectively. With computer science, computer engineering, and electrical engineering graduates all showing average starting salaries at about or above the $50,000 level, this disparity in favor of technology grads remains in sharp contrast to the under $30,000 average starting salaries of liberal arts graduates. In other words, liberal arts grads who never enjoyed the upside of inflated dot-com compensation packages to begin with, have seen their lower base pay eroded on average by a whooping 13.4%. Consequently, though technology grads have seen their total compensation lowered, they are still at the top of the pyramid when it comes to salaries, as you can see in Exhibit 5.2

Exhibit 5.2
2002 Summer Salary Survey

Average starting salary for new graduates	% Increase/Decrease in starting salary from 2001 to 2002		Discipline
$25,456	-12.80%		Psychology
$28,397	-13.40%		Political Science
$28,488	-9.50%		English
$31,201	+2.70%		History
$33,921	-9.40%		Management Training
$36,429	-5.30%		Business Administration
$38,459	+4.80%		Nursing
$39,405	-9.10%		Logistics/ Materials Management
$39,768	+ Less than 1%		Accounting
$39,953	-1.50%		Economics/Finance
$41,317	+1.70%		Civil Engineering
$42,705	-6.30%		MIS
$49,596	-5.90%		Computer Science
$50,123	-3.40%		Electrical engineering
$51,417	+0.70%		Chemical Engineering
$51,587	-4.30%		Computer Engineering

Source: NACE, National Assoc. Colleges and Employers, 2002 Summer Salary Survey

And there are other differences. For example, although the difference in the rate of unemployment between non-IT workers and people with IT careers are no longer quite as big, IT types generally enjoy rates of unemployment several percentage points *below* those of the general population. "During the recession of the early 1990's," says *Infoweek*, "IT unemployment averaged about 4 percentage points lower than the general jobless rate. This time, IT joblessness is about 2 percentage points lower."[12]

Finally, IT grads will probably see their unemployment pick up sooner than other segments of today's job market.   Therefore, if you are a technology graduate leaving school in year 2003 or later, your job outlook is probably on the upswing because of some anticipated new hiring. This hiring is probably due to pent-up demand within companies that have delayed projects to get through the past few years.

Not only will companies have fully amortized their capital spending for Y2K projects by the end of 2002, by 2003, companies will most likely be unable to further stave off the capital spending that gives them a competitive edge. Instead, says Michael Kelly, Chairman and Chief Architect of technology market research firm Techtel Corporation, by 2003, (barring some unforeseen calamity), two things will help stimulate hiring:

First companies will begin cautiously spending on low-end hardware such as PCs, notebooks, printers, and copiers as well as on their most critical applications, such as CRM, (customer relationship management). Second, other organizations will begin to accelerate their spending for new technology to retain their competitive edge. So, if you are finishing school in 2003 or later, you might see an easier job market than the past 24 months have shown.

Taking all of this data into account, you can enhance your career search by paying close attention to lessons from the recent past that identify pockets of industry, functional titles, and hiring retention criteria.

---

[12] "2001 IT Jobless Rate Was Highest Ever", Eric Chabrow, Infoweek, Jan. 11, 2002.

6. WHERE SHOULD I LOOK NOW?

In its May 2002 study called "Bouncing Back: Jobs, Skills and the Continuing Demand for IT Workers,"[13] the Information Technology Association of America, in a study with DICE, a well-known IT job board, produced technical job skills profiles across eight key IT areas (see Exhibit 6.1). This major study defined the IT job market and gave job seekers information on where to apply as well as how to survive economic downturns.

Exhibit 6.1
ITAA JOB CATEGORIES

Tech Support	Network Design/Administration
Programming/SW Engineer	Web Development/Administration
Database Development	Enterprise Systems
Tech Writing	Digital Media

Some of the interesting data that came out of ITAA's report:

- 92% of all IT workers work for non-IT companies
- Some 80% of all IT workers are employed by *small* non-IT companies
- Candidates with the greatest edge have expertise in at least one or more of these top areas: C++, Oracle, SQL, Java, and Windows NT, with some knowledge of security an important plus.

The study also showed that during the downturn people were more apt to keep their IT job if they worked at a non-IT company rather than an IT organization. This fact makes sense. The business of technology companies is to produce technology. But the past few years showed that no one was buying technology--many dot-coms failed, Y2K ended, and terrorism affected new project spending. The report concluded that as an IT

---

[13] Bouncing Back: Jobs, Skills and the Continuing Demand for IT Workers, May 2002, The Information Technology Association of America.

worker, you were more likely to keep your job if you worked at a non-technology-producing company whose basic businesses were still ongoing. In other words, if you were an IT worker inside a healthcare or insurance company, and your IT job supported an ongoing business, you were likely to keep your job.

Furthermore, the study identified jobs most vulnerable to cutbacks. Interestingly, it showed tech support positions and database developers as two professions hard hit by RIFs (reductions in force), although presumably for two entirely separate reasons. Database developers are highly paid and perhaps for that reason, among the hardest hit. Tech support is an area that rises and falls with a company's revenues. If a technology company sold its product to fewer customers than the year before, its revenues would fall and so would its customer base, requiring fewer workers.

Looking forward, ITAA came up with this analysis: Except for smaller non-IT companies, a group that in turbulent times tends to outsource more than to hire, "demand for IT workers has significantly increased year-to-year across the board . . . [Therefore] hiring managers in small organizations anticipate demand in 2002 to look much as it did in 2001. Those in medium and large organizations expect to fill jobs at respective rates of 2.5 and 3.2 times last year. The largest single year-to-year jump is within large non-IT companies, where demand jumps by a factor of 5.6."[14]

ITAA identified five types of positions as most likely to increase:

1. network design and administration
2. programming and software engineering
3. Web development
4. database development and administration
5. tech support

Tech support, which is expected to increase, will do so at 2000 levels, which ITAA calls the "high water mark year for IT demand."

---

[14] Ibid, p. 17.

Perhaps you're wondering why security isn't right up there among the top five job titles. Although security is a hot topic today, be aware that, while critically important, security is also a function that many organizations add to other functional titles and then "matrix manage" into the organization rather than create a job category of its own.   Though not included as a separate category among its top eight functional areas, ITAA says that if you are a security professional, "the most fertile ground for job seeking is large (1,000 or more employees) IT companies, which average 36.3 positions per firm, more than four times the security workers in similar-sized non-IT companies."

Now, if you are wondering where to job hunt by geography, you may want to note that on a geographical basis, the ITAA study identified the South as the area of the country holding the greatest number of IT workers (34%), followed by the Midwest (29%), West (22%) and Northeast (15%). The study said the South and West were the areas in which large IT employers were most likely to create IT positions. The Midwest was the area most likely for non-IT companies to create jobs. However, note that job seekers have also found luck outside of major metropolitan areas, where there are fewer jobs, but also fewer candidates to compete with.

Next, while ITAA gauges *previous experience* as the "single most important skill credential for obtaining a new job,"[15] and *education* particularly important for database developers, programmers, software engineers, enterprise systems integrators, and technical writers, *certifications* are more important than ever. They are, in addition to education, often critical to differentiating one candidate from another to potential employers.

In fact, ITAA looked at some 30,000 jobs posted on the DICE job board, (www.dice.com), and eight NWCET (National Workforce Center for Emerging Technology) career clusters covering tech support, database development and administration, programming/software engineering, Web development, network design and administration, enterprise systems integration, digital media expertise, and tech writing. The study concluded that "college degrees were the tickets to obtaining positions in software engineering and programming for recent hires,[16] with certification

---

[15] Ibid, p. 6.
[16] Ibid, p. 30

in the category of network design and administration higher in this than for any other NWCET job category as a credential carried by most recent hires."[17]

Sorting out all of this information, the new grad job market looks like this: New grads should find a warmer reception at medium to large <u>non</u>-IT companies, whose greatest hiring needs are in the job category of "programmer." Should grads prefer to work for a technology organization, be aware that this kind of organization is apt to hire *mid-level, experienced people*, mostly in the area of network administration, preferring new hires have certifications.

Next, should you decide that as a new grad, your best chances for employment come from taking on a tech support role and then moving up from there, ITAA warns that here you may be the most vulnerable to layoffs since your job depends on revenues. If you work in tech support for a technology company, you might find it easier to enter the market, but you might also be more vulnerable to layoffs if your company doesn't meet its revenue projections.

Last, certifications help not only in the job-search process, but also if you want to stay employed and move up. Exhibit 6.2 shows the types of degrees that are helpful in both IT and non-IT companies.

---

[17] Ibid, p. 33

Exhibit 6.2
IT WORKERS IN DIFFERENT ENVIRONMENTS

	IT/Technology Companies		Non-IT Technology Companies
Focus of Company	Technology/Computer Science Degree		IT + Business
Type of Degree	Technology/Computer Science Degree		MIS
Focus of work	Technical development, new products, technology advances		Business, working with established packages to make the business run
Layoff rate of IT workers during downturn	8%		4%
Characteristics of workers	Technical innovation, creativity, original thinking, fast growth, add technology to sell to customers		Understand the ROI of different choices, package and vendor selection, understand platforms (Oracle, Sun, .NET), understand business end-users and customers who need technology to work for them
Application development cycle	Develop, test; this is a product going out to customers that the company must enhance and support		Internal applications are often a shorter development cycle and used immediately
	Technology is the business		Technology adds to the business

Yet another look at the IT job picture showed interesting statistics on layoffs according to functional area. (See Exhibit 6.3.) In other words, if you were coming out of school with a specialization in telecommunications, the technology industry segment hit hardest by layoffs, you would probably be struggling to find work, at least for now.

Exhibit 6.3
JOB CUT ANNOUNCEMENTS
TECHNOLOGY SECTOR
2001 and 2002

Technology-Sector Job Cuts[18]

	January – April, 2002	January – April, 2001
Telecommunications	120,698	91,799
Computer	21,640	53,774
Electronics	14,839	46,118
E-Commerce	1,572	39,779
TOTAL	**158,749***	**231,470****

*29%OF ALL JOB CUTS ANNOUNCED THROUGH April, 2002 (555,783)
**40% of all job cuts announced through April, 2001 (572,370)
Source: Challenger, Gray & Christmas, Inc.©

(continued)

---

[18] 2002 Tech Jobs News Good, Bad, John Challenger, Challenger, Gray & Christmas, Inc., May 5, 2002.

(continued)

2001 Technology-Sector Job Cuts

	2001
Telecommunications	317,777
Computer	168,395
Electronics	152,882
E-Commerce	56,527
TOTAL	**695,581***

*36% of all job cuts announced in 2001 (1,956,876)
Source: Challenger, Gray & Christmas, Inc. ©

CASE STUDY:

Ed Reyes, a math and computer science 2001 graduate
From Rensselaer Polytechnic Institute
Now employed by a financial software firm as a
software engineer working in Visual Basic applications to
execute trades.

"I was a math and computer science major at RPI who graduated in 2001 with a Bachelor of Science and an overall grade point average of 3.8, including a GPA of 3.73 in my major. When I started interviewing, I began looking at business consulting with Internet firms, I also looked at an artificial intelligence group. Then I redirected my search to finance firms as I went along and my interest changed because of some classes I took.

I actually started job hunting in the first semester of my last year, in the September/ October timeframe, and began interviewing in earnest in November and December. During the first half of my job search I interviewed exclusively through the career center at RPI. I also used jobtrak (now monstertrak), and went to campus career fairs where I interviewed with IBM Global Services and some other larger companies.

Additionally, I used a headhunter available through the Web at CollegeHire.com, which is now defunct, but which specialized in post-college and technical jobs. Through them, I interviewed with companies including Enron, an AI company in Austin, and my current company, all of who were their clients. While the service was free to me, the job seeker, it cost the hiring companies a percentage of the candidate's salary.

As I went along I could see that I didn't want to be in a large company. I preferred a mid-size, 100-person organization with little bureaucracy – one in which I would have responsibilities and opportunities. I also began to get a sense of the different kinds of companies and their personalities, as well as the types of people they were looking for.

There were companies, such as Trilogy and Microsoft, that were not particularly seeking technical skills but were more interested in people with high problem-solving skills. My interviews at these companies meant talking to four or five different people, each one of whom gave me a different set of problems to work on. Some problems were computer-oriented; some were abstract and typical conundrums and brainteasers. At one company I took a standardized type of test and I had some 30 or 60 minutes to look at 50 problems. Some problems were quite abstract, testing my pattern-matching skills.

Usually, my interview sessions included three to six people and typically went like this: First, I met a human resources person. Then I met the engineers who did technical interviews and were working in the same technology I was working in-- Java, C, and C++. I sometimes met with a project manager, too. In one company, I met a vice president, although this was not a technical interview but instead a meeting to show they were a small company where bureaucracy was not an issue. Usually the human resources people didn't prepare me for the people I would be meeting, and I just assumed they would be technical people.

On the surface, I was asked a lot of questions that looked like math problems, but these were really brainteasers to assess my intelligence and whether I could think on my feet. In this part of the interview, the interviewers would give me problems to solve and then come back in 15 minutes and ask me to go through my solution out loud in front of them to see how my mind worked. When I interviewed, I definitely had to sell myself. I did a lot of homework about the companies I interviewed with, although with the smaller, private companies, it's a little hard to find extensive information. I also focused on each company's philosophy and the people who worked there. In one instance there was an RPI alumnus working at the company who was part of my interview process. I could see that it would have been valuable to have spoken with him to research the company before I actually interviewed.

I must admit that when it came to compensation, I didn't want to overpromise because I wanted to start low and prove my worth and then negotiate from there. But I also negotiated a six-month review, which was very positive. The company asked me, as well as my peers, to provide opinions about the work I was doing and where I might need to focus.

I certainly understand that it's a year later, and now the job market is definitely even harder. Therefore, to be hired, you must be really good and you must work even harder to be an impressive candidate because the talent pool is a lot larger and companies can afford to be more selective.

If I were working to find a job now, these are some of the things I would do. First, I would try to stay in closer touch with people from my internships from previous summers and also to think more about developing a clearer picture of what I'd like my career to be. Then, I would interview with more companies, including international organizations, and speak with more alumni.

What I can say now about interviewing is this: you must be prepared to show that you can solve computer programming problems and demonstrate your coding ability and be able to come up with code pretty quickly. To prepare, you should go over your computer textbooks and solve algorithm problems, working on the tricky problems that tend towards math. You also definitely need to be able to discuss any previous projects that you've done, and you will be asked to describe succinctly what you've done on a project, the problems you've had when you've been working on a project, and whether you could solve a particular company's technical problems. I can't emphasize enough that companies want to see that you are a problem solver. Therefore, remember that all your answers should be framed in blinking neon that shouts that you are a problem solver who likes to solve problems.

7. IDENTIFYING INDUSTRY AREAS OF SPECIALIZATION

Although nothing is for certain and no industry segment is invincible, it does help a jobseeker to identify industry areas and functional titles (programmer, customer support, business analyst) that are hiring. "If you were in wireless, telecom, or finance," says Dr. James Canton, Founder, Chairman and CEO of the Institute for Global Futures, a San Francisco-based technology think tank, "you must look at industries that are sunrise, rather than sunset and then figure out how you add value." This may mean identifying opportunities in healthcare or health informatics, nanotechnology, molecular computing, life sciences, defense, and homeland security. However, moving on to the next hot technology can also open up a Pandora's box of problems. In today's roller coaster markets, often, with almost no warning, today's panacea can quickly become tomorrow's problem-child. Market segments such as telecom can quickly fall out of favor for a variety of reasons. Nonetheless, if you find a technology that morphs into an enduring platform, pay attention to details.

One such area is CRM, customer relationship management. While CRM is no longer viewed as the surefire way to generate strong e-commerce revenues, it still maintains great promise for future growth. "While CRM spending will be down," advises Paul Greenberg, author of *CRM at the Speed of Light*[19] and executive vice president of Life-Wire, a CRM consulting firm, "it is still one of the more promising domains for IT jobs." In fact, while Greenberg says that most analysts believe that CRM software licenses will remain flat in 2002, they still anticipate a 12% compound growth rate in software product sales going from 2003 to 2005, with an 18% - 19% per-year growth of CRM services. However, much of that growth is happening outside of traditional commercial establishments. "The federal government and the public sector will probably be the most active solicitor of CRM over the next several years," says Greenberg.

Understanding that new grads in the CRM market are competing with an experienced candidate pool available at reasonable prices, not to mention that CRM is a business strategy

---

[19] CRM at the Speed of Light, Paul Greenberg, Osborne/McGraw Hill, 2001

not typically trusted to 21-year-olds, how can you still figure a way in? Exhibit 7.1 offers Greenberg's scenario.

Exhibit 7-1
ENTERING CRM

If you are working inside a company and they need a call center implementation using Clarify, for example, then you must understand how a call center cue works, how you use algorithms for peak volume times, and how this correlates to staffing for peak times. Plus you need to understand the processes within the company, including how tickets are filled out to identify and resolve problems, and how to create a marketing campaign.

Therefore a new grad with academic credentials for CRM will essentially be an apprentice, being paid a reasonable salary. In addition to needing some knowledge of the technical tools, new grads will probably start to work as an analyst figuring out how a business works, identifying the CRM processes that have to be changed, or modified. Then, they will need certification from a vendor that has a good shot in the market, whether it's Siebel, SAP, or Peoplesoft.

Paul Greenberg, Executive Vice President Live-Wire

8. UP IS NOT THE ONLY WAY

If by now your hopes of getting work within the commercial IT world are just about exhausted, there are a couple of markets that deserve a closer look -- the federal government and education. Both of these markets need candidates with "fresh" thinking and creativity. They are also markets that have funding and are expanding their use of technology in exciting ways. The federal government is under enormous pressure to bring its systems and databases up to date fairly quickly. The passage of the No Child Left Behind Act enacted by President Bush dramatically increased the funding for education, including technology initiatives. If you haven't thought about either of these markets, now is the time.

Education

In case you haven't noticed, public school systems are trying to recruit teachers with specific competencies in science, math, and computer science. There are programs such as Teach for America (www.teachforamerica.org) that offer new graduates scholarships, teacher training, and apprenticeships. Recent grads can spend a few years teaching until the commercial market thaws.

Also, technology products play a part in education. You, as an IT, computer science, or Web development enthusiast can appreciate that! Across the country, schools have realized that not only must our public k-12 educational institutions graduate students with the technical skills for 21st century jobs, but also that by using technology, teachers become better teachers, and students become more interested in their studies. There is growing momentum among high school, and even younger students, to learn tools such as PhotoShop, DreamWeaver, and Flash. Students want to build Web sites to show what they have learned about the environment, science, and history -- even help produce professional-level Web sites for non-profits. Whether a student is part of Mt. Diablo's Digital Safari Multimedia Academy showing her economics teacher how she would start a Web-based business or a student at Richmond High School in Richmond, CA building an award-winning robot in partnership with Industrial Light and Magic and then spending the summer on scholarship at Carnegie Mellon's summer program headquartered at NASA, the latest technologies are being absorbed into our schools at an accelerating pace.

"There are tremendous opportunities for IT people to come into education, particularly IT people with a computer science or computer engineering master's degree or PhD," notes Paul Kim, Ph.D., Chief Technology Officer at Stanford University's School of education. These opportunities include providing laptops for teachers and students, creating district-wide school and home Internet access, implementing personalized 24-hour accessible distance learning initiatives, going "wireless," or by providing parents with online access to their children's grades and attendance records.

There is also an incredible interest in Web development, and in all sorts of online applications to improve student and teacher training, not to mention the traditional IT functions. In fact, in a district whose budget is some $300 million dollars, you might find an IT department as large as 100 people with significant funding for salaries, hardware, software, and services. Furthermore, other kinds of educational institutions have similar needs. These institutions include technical schools, community and four-year colleges, universities, and private schools.

Consequently, educational computing is an area with multiple types of openings. You might decide to develop an online curriculum, sign up for a "crossover" graduate program merging technology and education courses, or use your telecom and networking backgrounds to develop an e-learning curriculum complete with Web-based courses and collaboration tools. With opportunities to grow into management as the chief technology officer of an educational institution that just might be offering its courseware online, working in this area at the manager and director level can be highly lucrative. "Online curriculum developers might make $80,000 - $90,000," says Kim, "with managers and directors earning significantly more."

Should you be interested in pursuing these kinds of opportunities, be prepared to investigate a wide variety of sites for specific IT in education job postings.   School districts don't always post in the traditional spots you have been investigating like monster.com and hotjobs.com (although many do post on craigslist.com, DICE, higheredjobs.com and the Chronicle of Higher Education's site). Once you find a job of interest, you might have to fill out and mail in a paper form.

Rest assured, however, that these jobs can be really worthwhile.

Getting a Job with the Federal Government

The federal government is on a hiring binge. In fact, the U.S. government employs 2% of the entire civilian work force, which represents 2,704,000 people earning an average of $51,000 per year, with some 305,452 new people hired annually on average. Roughly 28% of the federal workforce is eligible to retire, which amounts to some 700,000 people. Plus, with the attacks of 9-11, many agencies are building up their IT work force so they can counter terrorist threats, notes Dennis Damp, author of *The Book of US Government Jobs*.[20] There is work with the FBI, the CIA, the Transportation Security Administration, and other organizations that form the group of 153 government agencies with current openings. Therefore, if you are interested in finding *all* of the jobs, you will need to visit each government agency individually since the government abolished the centralized Federal Register. Although many openings are listed at the Government's Office of Personnel Management site, (usajobs.opm.gov), be prepared to hunt down all of the agency links, a task you can accomplish at www.federal.jobs.net.

With some 100 out of a total of 800 IT postings inviting new graduates to apply at the main Office of Personnel Management site, Damp concludes that as of July 30, 2002, all 153 agencies together probably have three times that number of new jobs for graduates. Pay scales range anywhere from $22,000 to $39,000 plus location pay, which can increase that base number by 10% to 15%.

New hires need either a bachelor's degree, three years of general experience, or one year of professional experience to come into the system with a grade level of GS-5 (general schedule) to GS8. Grade levels and pay ranges go up with two

---

[20] The Book of US Government Jobs, 8th edition, Dennis V. Damp, Bookhaven Press LLC.

years of graduate work or a master's degree. Ph.D.s come into the system at the GS-11 rank.

The trick, however, to successfully applying for a government job is to first know where to find these jobs, and then take the time to apply in the specific format that the agency requires. Plus, you would be even better off with some "inside information" or personal contact to ease your way through the federal bureaucracy.

With just 20% of all jobs in the federal sector requiring exams, it's wise to note that when it comes to the government, the exam is the application itself. Therefore, says Damp, you must do a really dynamite job in preparing your resume, which requires completing 43 data elements in your application that are rated against certain standards.

Part of the fun in applying for a job with the federal government is that the application process itself shows you how these agencies will work once you're inside. Be prepared for a labor-intensive process. First, you must submit your resume in a specific format, or you can submit the OF-612 form "Optional Application for Federal Employment," or a Resumix version of your resume. Luckily, you can buy Datatech software, the Quick and Easy Federal Jobs Kit Version 6.5 for $49.95, available either toll free at 800-782-7424 or www.federaljobs.net. The software converts your resume from one approved format to another. Moreover, you will also need to fill out OPM Form 1203, Form C, a form that is scanned, allowing someone who might not know anything about the job to rank your qualifications.

According to Damp, "you could actually bid on the same job series with two different agencies and the qualification questions will be different."[21] So if you apply for a Federal Job and apply often, as Damp advises, be prepared to submit individualized paperwork each time.

In addition to your resume and Form C, government agencies may require you to fill out two sets of paperwork. These sets are known as KSA's (Knowledge, Skills and Abilities) or KSAO's (Knowledge, Skills, Abilities and Other Characteristics), and QRF's, (Quality Ranking Factors). "You have to remember

---

[21] p. 109

this is the federal government," says Damp. "You have to complete the paperwork to beat out your competition." Therefore, many jobs require you to submit one page per KSA or KSAO to complete your application, plus any paperwork or tests individual agencies or jobs might require.

In addition to applying for a federal job through this process, you have other ways to get familiar with how government structure works, while developing a network of people to vouch for you. If you are comfortable with a tour of military duty, you can join the ROTC while you are in college or once you are out, apply to Officer Candidate School. "If you're interested in government, you must make a conscious decision to use your skill set in government," advises Adrian Barbour, Western Regional Recruiting Manger of Anteon Corp. Plus you must understand government culture and get security clearance, which might be best gained through military service. "If you are part of the ROTC and go through a tour of duty, you will have four years of experience dealing with the Department of Defense and experience dealing with federal qualifications and things the federal government looks at. Plus you will understand how the government works and know people in the system who can vouch for you."

Though many IT graduates think that military service refers only to the Army, Navy, or the Air Force, heed Barbour's advice not to overlook such agencies as the U.S. Merchant Marines and the Coast Guard. These agencies deal with federal acquisition regulations and are hiring IT people to modernize their legacy systems, converting them to TCP/IP and Web-based servers.

Exhibit 8.1
FIRSTGOV.COM

Source: www.firstgov.com

To figure out whether a military or quasi-military opportunity is for you, start reading *Federal Computer Week Magazine*, www.fcw.com. Other publications discuss government technology initiatives, including: *Federal Headline News* and *FSI Weekly*, http://www.fedsources.com/, which also contains information about state and local government technology initiatives; http://www.fedsources.com/elements/index/news/hn-sl.asp, and defenselink.com, with its e-business initiatives, described at http://www.defenselink.mil/acq/ebusiness/. Additional information about state and local government technology initiatives are listed at on the Government Technology site, http://www.govtech.net/. Exhibit 8.1 shows www.firstgov.com.

Military.com, a job site powered by DICE, is for the recent college grad. There are also associations to check out, such as AFCEA (Armed Forces Communications and Electronics Association at www.afcea.org), and the NDIA (National Defense

Industrial Association, www.ndia.org), which have local chapters throughout the country.

9. NETWORKING: YOUR MOST IMPORTANT TASK

As you start your job search, you'll be spending time at your college or university's career counseling office. You will probably be surfing a number of Internet sites. Some good sites are Monster (www.monster.com), hotjobs (www.hotjobs.com), brassring.com, techies.com, careerbuilder (www.careerbuilder.com) and flipdog (www.flipdog.com). On these sites, you can search for jobs and post your resume. But, no matter how much time you spend on the Internet, it will not be enough.

First of all, even among the Fortune 500, Internet postings represent just a small portion of the actual jobs that are available at any one time. "Centralized sources such as Monster are great to see who is hiring," comments Yves Lermusiaux, founder and president of iLogos, the research division of Recruitsoft. " But in reality, the ratio is three to one, which means for every job posted on Monster, at least among the Fortune 500, there were three jobs posted on the company's own Web site."

Second, even if you see a job you want posted online, your chance of becoming a candidate for the job increases dramatically if you apply to the company's Web site rather than through a portal (you can do both). The reason for this is "time." Recruiters need to log on to external career portal sites separately, an extra step to searching the resume database they have for people who apply directly through the company's own Web site.   So, even if you find your favorite job listed on both sites, statistically you can increase your chances by applying to the company's own database rather than through the external site.

Third, even if you apply through the company's site, but not to a specific person, this still won't be enough for the hiring manager to know you're alive. The way to "beat" the system is to find the right person within the company and let that person know that you are an available candidate with just the right skills and personality.   Then you will truly have the search process on your side.

Therefore, no matter the power of the Internet or the connections you have through your family, pay attention to the words of Susan RoAne, the author of *How To Work a Room*[22] and *The Secrets of Savvy Networking*[23] whose two axioms you should live by:

"Networking is not just for when you need a job, it's for all the time."
And
"Networking is not about who you know; it's about who knows you!"
Source: Susan RoAne, author

Employee referrals are the major way companies hire. They often account for as much as 50 percent of an organization's new hires. Your problem is that when you are at school, you usually don't know all that many hiring managers except for your parents and their friends, and friends of their friends. Consequently, now is the time, with your almost newly minted degree, to start expanding your list of contacts.

Because many companies are cutting back their travel budgets, campus recruitment, and corporate internship programs, many college career centers are no longer full with corporations lining up for their technology grads. In fact, company visits for campus recruitment were down in 2002 by 50 percent.

Although the college career center will undoubtedly play a large role in your job search, particularly at the beginning, it is but a mere prelude to the greater search efforts today's successful grads are making on their own. These efforts include extensive networking with alumni, interacting with professors, plus a lot more.

On some campuses, students have set up travel and bus trips to bring graduating students to the doorsteps of the companies they want to work for. Other universities have established student-run BLPs (Business Leadership Programs) and seminars. Here, the Association of Students at schools such as Harvard or Stanford form a Student Enterprise unit as part of the Student Association, and then hire a paid student coordinator to go out and actively solicit corporate sponsors to come onto

---

[22] How to Work a Room, Susan RoAne, Warner Books, 1988.
[23] The Secrets of Savvy Networking, Susan RoAne, Warner Books, 1993.

campus and look at this year's grads. "I started working in January and kept working right on up until the following September when the students who were participating became seniors," says Stephanie Chang, Stanford University's Business Leadership Program Manager.

This is how the program works: At Stanford University in Palo Alto, CA, which accepted some 65 students, BLP is a four-day event with daytime seminars followed by sponsored dinners each evening. Daytime topics include academic and practical insights about business from Stanford's Graduate School of Business faculty, plus preparation from the Career Center and Oral Communications department regarding how to do well at interviews. Each night there is a recruiting dinner sponsored by companies the program has enticed into attending. "A meaningful number of students have gotten jobs out of this," says Ms. Chang.

But whatever formal and informal programs exist on campus, your major focus during the job hunt should be on developing a list of specific people to contact and then figuring out ways to contact them personally. Since 2000, the dollar value of MBAs from great schools may have plummeted right along with your NASDAQ stock.[24] Now it's time for IT/tech types to emulate the MBA's bloodhound instinct for power networking. Generate a list of appropriate people who want to help you and then work your way into their comfort zone.

Such networking includes staying in touch with anyone you have ever worked for and interned with. It includes talking with the professors who have taught you, speakers who have presented to you at conferences and seminars, and students who have graduated not only from your particular school but also who have graduated from your school's business, law, and other schools. You should seek out those who have gone into business for themselves, those who have become the heads of hospitals, gone into educational or scientific computing, or even joined the military. In other words, you want to go far beyond people who were in your department at your school to people who were in any department of the school or graduate school you've attended. You want to seek out contacts who work at corporations, in

---

[24] "What's An MBA Really Worth? A Lot Less Than You Think," *Business 2.0*, July 2002.

government, and in other organizations that are - or might be – interested in hiring a "you."

"The largest, best method of companies hiring to bring in new people is through employee referrals, whether it is your mother bringing in your resume or a friend of an acquaintance. This is not going to change no matter what the technology," notes Mark Mehler, co-author of *CareerXroads.*[25]

So, after visiting the career center and meeting with the grads who have volunteered to counsel new grads, take yourself over to the college's Alumni Office. See if there's any recent or older grad who might live or work in the cities you're targeting and who has become an active donor to the university. These are the alumni who will usually help you.

Once you have compiled your list of contacts, it's time to prepare yourself before you call them. First, you will need to rehearse what you will say about yourself and what you will ask for, and next, you must research who they are exactly and why you are calling them. It's just not enough to say, "Hello, Mr. Jones. This is Hank Smith. I'm graduating from XYZ University with a degree in IT and I need a job."

First of all, if you call Mr. Jones and ask for a job, he probably has been getting 50 calls from other people on the same mission. But, if Mr. Jones has become an experienced manager at a company you're interested in working for – or if Mr. Jones is an experienced IT manager, even if he is not hiring now -- he can still help you prepare for interviews in a way that will benefit you. Just don't approach him cold. Instead, research who he is. Use the Web and, if necessary, visit the college library.

"If you don't know what you want to do and need a focus, college career centers are useful," says Margaret Dikel, a librarian turned author whose expertise is in career sites and the job-search process. "They can give you a list of employers who recruit on campus and a list of the alumni who work for various employers so you can network. But every aspect of what the career center is telling you requires research, and for 60% of it, you'll need to go to the college librarian."

---

[25] CareerXroads 2002, 8th edition, Mark Mehler and Gerry Crispin, MMC LLC.

You'll probably start out by looking in the *Occupational Outlook Handbook*, and after that, in trade journals. You'll want to know everything you can about a target industry and particular companies, both the employers you want to work for, as well as their competitors. And if you can, you'll want to research the specific technology and business initiatives these companies are launching, the products and platforms they are using, and their major customer wins and losses. All of this leads to one key point: when you call Mr. Jones, you can say something meaningful that will keep Mr. Jones on the phone speaking with you. It will also help him to remember you, particularly when there's an opening.

Start out by introducing yourself -- your name, school, how you got his name, and, of course, asking if this is a good time to talk. While 99% of the people you call will want to help you, should you hit someone who is too busy, not interested, or just having a bad day, don't forget, you can always just hang up. Then compose yourself, redial the number, and say, "I'm sorry. We must have been cut off."

Now you have researched Mr. Jones sufficiently and know what he does. You also have asked whether he has a minute and isn't in a meeting and can take your call. Even if Mr. Jones says his company is not hiring now, he can still help you. Mr. Jones works at a financial services company that has a CRM initiative, a fact that you know having looked at their Web site. You next convey your research.

YOU SAY: "I want to start by saying that I'm most interested in working for a financial services company that is focused on a CRM initiative, and I want to be well-prepared for my interviews. I am just beginning to interview and it would help me to know, as a new graduate, if I need experience with a particular package or if there is any specific coursework I should emphasize." (Allow time for Mr. Jones to answer, and don't interrupt him.)

YOU CONTINUE: "I also have another question, regarding whether I should start working on a certification, if that would impress a hiring manager, or if my degree with a major in computer science is enough."

YOU: "And finally, my last question is this. The job market is so tight. I really want to work in software development but I'm thinking of interviewing for a tech support role just to have an entry point. Do you think this is a good strategy?"

Of course, you can also ask Mr. Jones if he has any thoughts about companies that are hiring, whether they are in the same or a different area, or of any Web sites or associations that are particularly meaningful. But above all, don't fail to contact him and thank him. E-mail is okay. Just write the subject line as: "Thank you from Jeffrey/JanetJobseeker" so he knows that you've followed up.

Now presuming you are running out of contacts or just like using the Web, there are a few networking sites you can try where you can expand your network without staying online in time-consuming chat rooms. You might try:

- www.alumniconnections.com/your university name goes here (For example www.alumniconnections.com/yale)
- www.datanet.com,
- yahoogroups
- www.ryze.org

Additionally, there are technology special interest groups, association meetings, networking events, chamber of commerce meetings, technology conferences, and meeting calendars that are generally listed by area of the country or industry specialty. For example, New York has www.alleyevents.com, whose extensive recent postings included the following:

```
Tuesday, May 14, 6:30 pm
Silicon Alley Entrepreneurs Club (SAEC)- CEO Meet
Organizer: Silicon Alley Entrepreneur's Club
Location: Zanzibar
Directions: 645 9th Avenue
(Corner of 45th St)
NY, New York
Cost:
More details:
http://www.alleyevent.com/ae/ae.nsf/1/6EAA

Tuesday, May 14, 6:00 am
NYNMA's CyberSuds Networking Event
Organizer: New York New Media Association
Location: One 51, 151 East 50th Street
Directions:
```

Cost: Non-members $15 in advance, $20 at the door
(NYNMA members, free)
More details:
http://www.alleyevent.com/ae/ae.nsf/l/6EF6

Wednesday, May 15, 6:00 pm
Venture Economy Forum -- Shaking the Money Tree
Organizer: NY eCommerce Association
Location: Harvard Club of New York
Directions: 27 West 44th St
Cost: NYeComm Members: $35/Non-member: $65
More details:
http://www.alleyevent.com/ae/ae.nsf/l/6EB2

Wednesday, May 15, 7:30 am
What VCs Want Today
Organizer: iBreakfast
Location: Masons Club
Directions: 71 W. 23rd St @ 6th Ave. 2nd Fl.
Cost: $50 Non-members $60 Add'l $10 at door
More details:
http://www.alleyevent.com/ae/ae.nsf/l/6EDA

"Source: Lee Hecht Harrison and AlleyEvent" Copyright
AlleyEvents, Inc. 2001

   If you are wondering whether any of this pays off, take
heart. It does.

CASE STUDY:
Leeann Sobehart
Business and marketing graduate, May 2002
The University of Dayton
Now employed by the University of Pittsburgh as a
Communications Specialist

"When I started searching for a job after graduation, I was primarily looking for an advertising job that focused on technology companies, one that was within an advertising agency or corporate advertising department.   While I had used my college career center to find an internship, I knew I wanted to come back to my hometown to work. Therefore, I expanded my strategy to focus on the Internet, since my college career center really didn't have leads for cities that far away.

I started sending out a lot of resumes but rarely got a response. When I started using the Internet, I simply logged on every day. While I had accounts with Monster and Flip Dog to push appropriate jobs to me, I soon saw I wasn't being e-mailed a lot of jobs and that perhaps only two or three employers picked up on my resume once it was posted.

So I started to use the Internet in another way. I began using career sites to do a variety of job searches, rather than cold call companies, using the sites to see what was available. Once I could see if a company was hiring, I could find a person to contact, and then e-mail my cover letter and resume directly to this individual. Then I generally waited to hear, although in a couple of instances I was very, very persistent and followed up with phone calls.

I happened to get my lead for the university job I have now when I was home from school. Someone that I had worked for, who happened to work at the university and happened to remember I had just graduated with a joint business and marketing degree, had the department contact me. I knew there would have been a lot of competition for the position but I luckily found them at just the right time. They hadn't really started a search yet and the woman whose position I was interviewing for had just announced she was going to be leaving rather quickly, which meant the department needed someone who could take over immediately. During my interviews, I was able to say that I could start at once and meet with the person who was leaving and learn the routine before she left.   The interviewer just said, "Come on board."

This was almost unimaginable to me since I'd had such a frustrating search, including having been at home from school for more than six weeks with absolutely no leads. I actually felt I was getting increasingly desperate. I was doing temp work with a variety of agencies throughout the city at just $10 per hour, until this one particular day when I was offered my permanent job after turning down a position with a temp agency that very morning!

I also want to add that when I got my job, there was absolutely no negotiation regarding salary. I was offered the job at just under $25,000 and took it, understanding it had one great upside for me -- I have a chance to further my education at a major institution at a very reasonable cost.

There are a few tips that helped me that I'd like to pass on to other people.   I always followed a resume format that had been recommended by my teachers and guidance counselors. They all told me to keep my resume to just one page. Plus, I always changed the objective based on the position and the employer I was applying for.

My resume started with my education and academic honors, and detailed my work experience and extra-curricular activities. For most employers, my list of academic honors, my GPA, the fact that I had held an internship while being a full-time student, and that I was on the rowing team -- a varsity sport -- were the things that were most important to them.

My final advice to other people who are looking is simply this: While it is very tough to be out there looking for your first job, just keep on looking and stay positive. Don't get discouraged. Talk to a lot of people because people that you might not think could help you could wind up being extremely helpful. This includes friends of the family, current and former employers, really anyone.

And then we have Ryan Brateris's story.

CASE STUDY:
Ryan Brateris
Texas Tech University in Lubbock, Texas
May, 2001
Now working for a Texas electric and gas company

The technology job market was booming the year before I graduated. Businesses were salivating for young college graduates with a technology background and wanted us to help them run their businesses into the future. Money was already set aside for new hires and future projects. Companies overhired, expecting the market would go even higher.

I was one of those hired during that market. I was hired by one of the Big 5 (or should I say the Big 4 now) to work in their consulting firm. I was ecstatic because this was always what I wanted to do. I was to travel the country and even possibly the world, meeting new people, working with some of the newest technologies, working on multiple projects, which would help extend my knowledge and marketability -- not to mention that I would be getting paid! But then the market began eroding, slipping on a daily basis, never coming back.

In my last semester before graduation, once, late at night, I began talking with one of my good friends at school who was slightly older, with a family and kids. We were working at the computer on a project that needed to be completed by the end of the week, and started talking about the future and what it might bring. I remember that he asked: "What do you think is going to happen when all of those projects, allotted for the future, end? What will happen to all of those employees hired for those projects?"

We were optimistic and said: "Ahh. ... Something will come up!" But later on, I learned that my start date had changed from mid-July to late August, and I remembered that conversation. But still, I remained optimistic, thinking that this would blow over.

By the end of July, I was contacted again, telling me my late-August start-date had become mid-September. I saw the market falling, with no vision of recovery, and I began job hunting. Although, at the back of my mind, I still thought all I had to do was wait.

But then I was contacted again, this time learning that my September start-date was now mid-April, and that it no longer was a firm date, just a tentative date, which meant that in April the company would contact me to let me know what its plans were. By then I was shocked, but I felt I had what it took to get a job in the business world: I knew I was good with people, that I had my degree and basic technical knowledge, and that I was trainable. But then I started thinking, "These things don't really matter if there are no jobs out there to fill."

So I started networking with family members, although I had uncles who were getting laid off. When my relatives couldn't help me, I turned to a neighbor who was an important person in one of our biggest Dallas companies, an electric and gas supplier for Dallas. He had known me since childhood, and knew the kind of person I was. He took my resume and said he would pass it on to someone who could help me.

A week later, I got a call back from TXU to set up my interviews. I was so relieved, but not completely, because I knew this was only the beginning. I knew I didn't have the position just yet, but I was still optimistic. I knew I could get this job.

After a four-hour interview, I did get an offer, and I was tempted to play with it, but I quickly realized I had nothing to counter it with because of the market. So I accepted the job, and I was set. I had a permanent position as an associate programmer waiting for me in mid-September. I couldn't wait to start. I also wanted to confirm the faith my neighbor had in me.

Mid-September came along and I showed up for my first day at work, ready to get things rolling. I might have shown too much enthusiasm. I quickly saw I had to slow it down, taking time to get to know my coworkers, so I began to assist them in different tasks. I didn't exactly know yet what I would be doing or the kind of project I might be assigned to.

My first couple of months were mainly spent in training and learning about what was expected of me. In December, I was placed on a small team that was using a tool called Documentum to house the company's Web content. I was brought on to alleviate some of the easier tasks so that more experienced programmers could focus on the administrative and development side of Documentum. I made sure I learned quickly and did whatever they needed me to do.

I really liked the environment I was working in. I also liked that I was working with TXU's Internet Web site and its intranet also. I could see the tool was quite useful in a lot of ways and I wanted to learn more. I decided that I should show some initiative and learn more about Documentum and what it could do on my own. I started helping in all areas so I could get to know more about what my other team members were doing. I continued to learn and help out. In the end, I could actually tell someone what I do at work, which made me really happy.

I can say that people working in technology need to remember that IT departments exist in every company, but except for companies that are building technology, IT people are considered overhead. Thanks to my current experience, I also see a lot more clearly what technology means to a company's revenues.

Now, with all the layoffs, companies seem to ignore the fact that when there are so many cuts, there won't be enough personnel to upgrade to new systems and build new programs. Part of my solution is to spread the knowledge of one person to another and cross-train individuals so companies don't get pigeonholed with one person in the driver's seat.

10. INTERNSHIPS, CERTIFICATIONS, AND TESTING

In a market rich with talented, experienced candidates who are no longer upping the ante every six months, rest assured employers are spending neither time nor money to train or educate candidates if they don't have to, particularly since training budgets, along with company revenues, have shrunk substantially. So you may find yourself in the job market competing against someone who graduated two or three years ahead of you who has experience that you don't have, either in a particular functional role or with a particular tool.

There are, however, a couple of ways to level the playing field. One way is through an internship. The second way is by obtaining a certification or passing enough tests to show you have the practical knowledge and experience for a particular job.

Looking at some recent data from Eric Lochtefield, director of an internship program called the University of Dreams (www.uofdreams.com), the drop in companies offering internships has been enormous, with 70% of internships among the big technology companies totally disappearing between 2001 and 2002. "In 2000, Cisco alone had 1100 interns and offered them stock options," he states. Lochtefield attests to calling 600 companies to get 90 internships for the summer of 2002, many of which were paid for by the students themselves.

But still, working as an intern, even if for no money, is an enormous leg up on the job market. In the first place, a company provides you with real-world experience, which is invaluable. And second, because they know you, you have a way back into the organization once you've interned there for your first post-graduation job. "We have a Babson graduate who did an internship at USB Paine Webber, who stayed in touch with them over the winter and just was hired full-time," notes Lochtefield.

If you cannot get an internship, paid or otherwise, at the organization or type of organization you'd like to work for, then volunteer at a non-profit, viewing this experience as just another cost of your education. "Contact organizations that can't afford to hire someone like you and offer to do something for them," advises Margaret Dikel, author of the 2002-2003 Guide to Internet

Job Searching.[26] In other words, go to an animal shelter and create a Web site for them with photos of animals for adoption. Get some real-world experience any way you can so you have references from people in the work world who can speak up for you. But don't just pick any non-profit. Investigate board members *before you select* since non-profits usually have board members who are well connected and can become your post-internship advocates.

Yet another way to increase your visibility is through the certification process, through which you become trained and certified in a specific skill an employer is seeking. Understand, of course, that this strategy can vary greatly in terms of the cost of the courses and the certification tests. Also, different certification programs require you to be active in a field for a certain length of time before you can qualify for the tests.

Becoming a certified Microsoft MCSE might take seven months from start to finish, while it might take years of experience and coursework to become a certified Oracle DBA. Overall, the trend shows that increasingly, companies and organizations are not only focused on selecting candidates who have the right credentials so they are more productive more quickly, they are also using testing and certifications to filter through the candidates flooding into their resume databases. "If you're a clever wordsmith your resume might say you have four years of SAP experience and you'll be hired," notes David Foote. "But now certifications have become more popular because they are a normative measure."

The Office of Personnel Management recently required all candidates attending an online job fair to take and pass two tests supplied by Brainbench, a certification company which does 35% of all IT certifications and administers 1.5 million tests per year. IBM is one of many other organizations telling colleges that new graduates who are applying for programming jobs need to have the Brainbench Java 2 certification at the front end of the process. "The reality is that companies right now can insist on experience

---

[26] The Guide to Internet Job Searching, Margaret Riley Dikel and Frances E. Roehm, McGraw Hill/Public Library Association, 2002.

and that is what they are doing," says Mike Russiello, Brainbench CEO.

Therefore, new graduates may find themselves going to the Brainbench (or another screening vendor's) site, paying the fees per test, which vary from $49 on up per test, taking the test, and then setting up a public folder to let potential employers see their scores. "Candidates can choose to either post their scores in their public folder or else retake the test and possibly improve their score before making the data public," notes Russiello. Exhibit 10.1 shows some popular certifications and the time it takes to obtain them.

Exhibit 10.1

Sample Certifications	Years of Experience to Obtain
SAP R/3	3 - 4 years
Cisco Certified Internetworking Expert (CCIE)	10 years
Project Management Professional	4500 hours of project manager work over 3 years
Certificate for Information System Security Professional	3 years of IT security experience

Source: Computerworld, 7-29-2002[27]

---

[27] "Standing Out From the Crowd," by Mary Brandel, Computerworld, 7-29-2002.

There are other considerations to testing and certifications. Training and certifications represent good revenues to vendors, plus it is to the vendor's advantage to know that their customers and partners are "certified" up to particular standards. You can imagine that with training and certification dollars growing, a number of alternate sources of training and certifications have sprung into this market, giving you more choices.

Some vendors and certification companies will say that the value of in-person and online testing from the vendor itself, be it Oracle, Peoplesoft, or SAP, for example, is higher than that obtained through an independent training, testing, and certification organization. Part of the discussion centers around money. While vendor courses and tests may cost more, there seems to be more cachet from having an Oracle certification from Oracle rather than by becoming Oracle certified through an independent organization -- unless your scores are so high that they show a superior knowledge in a particular package or field.

In addition to being able to choose between the vendor and the independent training organization, you can also choose how you acquire the knowledge -- either through in-person or online courses, or even through self-study. You can select what suits your schedule and pocketbook best. "Our most popular class is an SQL class, which costs $2,000 in the classroom but $399 via self-study over a 3-month period with an online instructor who can help mentor the student," notes Chris Pirie, Vice President of Oracle University. Then there is the matter of the tests, which, in this example, cost $90 for the first test. However, as you climb the food chain, be prepared to spend more "to become an Oracle certified professional and reach that professional level. Here you must take a face-to-face class with Oracle and pass another two exams," notes Pirie.

But right now certain skills are holding up very well, even in today's constrained market. Per Pirie, an Oracle DBA might earn between $60,000 - $120,000, with the person holding the Oracle certification having as much as a 21% salary advantage. David Foote agrees. "Certifications account for an average of 8.3% of base pay and have lost 1-2% during the year while skills pay is down 18%," he notes. Plus, according to a recent salary survey in *Computer Reseller News* highlighting Cisco CCIE, Sun Java, IBM

Websphere, RedHat Linux, Oracle DBA, and HP OpenView certifications, some certification pay differentials depend on geography. "In 2001, West Coast technicians earned an average of $2,500, or 2.9 percent, more than those on the East Coast," the study noted.[28]

Additionally, certifications provide access to a network of companies and opportunities that you might not otherwise have access to. A company's partners often must hire certified professionals and therefore have Web sites to make these interactions possible. For instance, Microsoft, SAP, Siebel, Oracle, and PeopleSoft all have alliance partners and customers who use their software and services. After you are vendor-certified, you usually have access to this network. While software vendors don't act as job placement agencies, they do help refer people who have passed their exams into their technology network exchange. Through these exchanges, people looking for work can profile themselves and hopefully be found by companies answering RFPs and seeking consultants for either full-time or contract jobs.

---

[28] Computer Reseller News 2002 salary survey:
http://www.crn.com/sections/special/ssurvey/ssurvey02.asp?ArticleID=35
952

11. JUST A WORD ABOUT RESUMES

With so much already written on the topic, let me just give you a few tips as the final part of my coaching. If you need help writing a resume or creating an appropriate cover letter, there are some wonderful reference books and Web sites available with plenty of good resumes for you to look at. First of all, there's a resume tutorial at www.careerbabe.com, and you can also use Yana Parker's *Damn Good Resume Guide*[29] and Joyce Lain Kennedy's *Resumes for Dummies*[30], *Cover Letters for Dummies*[31] and *Job Interviews for Dummies.*[32] Then, there's Susan Ireland, in print with *The Complete Idiot's Guide to the Perfect Resume,*[33] and online at www.susanireland.com and Bill Frank's *200 Cover Letters for Job Seekers,*[34] also available online at http://www.careerlab.com/letters/default.htm.

Here are just a few words from Margaret Dikel and Susan Ireland to enhance what you already know about writing your resume. Before you send in a resume, think about how you want to present yourself in a particular situation, advises Ms. Dikel. And don't let your resume look like you can only do one thing. You must show that you have a number of skills that prove to the employer that you are worth more than the salary that you will be paid -- that you are a great bargain at this price. Personalize your resume every time you send it out and for every conversation you have. "Remember," cautions Dikel, "that the first scan of your resume is to make sure you match up. A:A and B:B. Make sure your resume presents your exact skills."

---

[29] The Damn Good Resume Guide, Yana Parker, 10 Speed Press, 1996.
[30] Resumes for Dummies, 3rd Edition, Joyce Lain Kennedy, IDG Books, 2000.
[31] Cover Letters for Dummies, 3rd Edition, Joyce Lain Kennedy, IDG Books, 2000.
[32] Job Interviews For Dummies®, 2nd Edition, Joyce Lain Kennedy, IDG Books, 2000.
[33] The Complete Idiot's Guide to the Perfect Resume, Susan Ireland, Alpha Books, 2000.
[34] 200 Letters for Job Hunters, William S. Frank, Ten Speed Press, 2001.

Additionally, advises Ms. Ireland, there are some tactical steps for you to take. If your counselor gives you a format or fonts to use, don't be creative. Use the format. The person looking through a bunch of resumes to find data can search them easily (and find your particular information) when they're in a standard format. Use a job objective statement. Also, put your title in bold, in a little bigger print size right in the center at the start of the resume. This treatment will put you on good footing to negotiate your job title and salary.

Finally, when you are creating an online resume, Ireland points out, you can put your name on the left hand side, which is what most people do, but if you are producing a hard copy resume, put your name in the center or the right hand side. This way, if the hard copy is bound into a book or put into a file folder, your name and contact information are readily visible. Putting your new degree or certification right next to your name -- "Chris Brown, MBA" or "Margaret Smith, Certified MCSE" -- may be just the ticket that puts you over the top.

12. HOW SHOULD YOU INTERVIEW?

No doubt, today's new grads need to be prepared both technically and personally for tough interviews. Such interviews might mean meeting many people in the evaluation process, coding on whiteboards, and answering questions about your work ethics or the hours you're willing to put in. Therefore, even though you're new at the game, once again, homework is in order.

First, remember that a company's strongest source of new hires is its own internal employee referral system. Therefore, your interviews will proceed a lot more smoothly if in fact someone who works at that company has recommended you as a candidate. Second, having an insider help you will also enable you to research the company a lot more fully and get the inside scoop.

When recruiters call to set up your interview, they might just say, "We'll see you here at 2 pm. Go to our Web site for directions." Therefore, it's up to you to ask: "How many people will I be meeting?" and "Can you tell me what they do?" By this, I mean what *each* person does.

On the technical side, prepare for your interviews by talking to people who have recently gotten jobs or to your contacts and mentors who are hiring managers themselves. You can also read *Ace the Technical Interview* by Michael Rothstein[35] to see how the technical side of an interview might be handled. Most likely, the company will ask you to solve a particular kind of problem. Remember that what is important might not be the exact solution you come up with but rather the logic you use to get there.

Many technical companies have downloadable toolkits available on the Web. If you're interviewing with a company that has a particular software solution, download its tool kit and begin working with it – and perhaps those of a competitor or two -- to show that you have both initiative and some familiarity with the toolset that your potential employer is selling or using.

Next, work on your attitude and presentation skills. Often, decisions are made within the first few minutes of an interview. In

---

[35] Ace the Technical Interview, 4th edition Michael Rothstein, McGraw-Hill.

your first job, what you want to convey most is that you have an intense passion for ferreting out information and working through problems on your own, that you are flexible and also a fast learner with real initiative. Plus, if you don't know something, you should not only say, "I don't know," but also follow up with what you would do to find out. Exhibit 12.1 offers some more advice.

Exhibit 12.1
INTERVIEWING ADVICE

1. Do your homework. People must know about the company and the job they're applying for. If you can, find contacts within the company and learn about the environment from someone on the inside.

2. Understand the technology. If you are interviewing for an analyst role working in a PeopleSoft environment, then you must know what PeopleSoft is. Is there a package in the environment that is the standard for your work? If you are interviewing for a job as a data analyst, how well do you understand the theory behind what you will be doing and when to compromise that theory?

3. Demonstrate a positive attitude. The best asset a person right out of school has is that they are in learning mode - it is easy for them to learn. They can fit that new knowledge into the way things are done in that environment.

Loretta Smith, Senior IT Consultant, T. Rowe Price

Finally, when it comes to getting an offer, remember, it's a buyer's market. Do your homework so you know what other grads in your market are getting paid. Be prepared to relocate on your own dime if you have to. "Keep a grip on your ego, and don't go in there with a set number," says Silicon Valley recruiter Jeff Banks. "Don't pitch multiple competing offers. Simply say: 'I want you to make me your best offer and compensate me the way you compensate other people who do similar things.' "

Then, when they offer you the job, take it and remember to show up for your first day of work at the appointed time. If anything else comes through, don't just not show up, which happens all too frequently. Call the company and explain what is going on.

13. WHAT TO DO IF YOUR SEARCH STALLS

Remember, today's job market is tougher than ever. It may take you a while to get interviews, never mind new work. Therefore, once you're out on the job market, possibly living at home while you search, it is imperative to keep broadening your network, while still finding the internal stamina to keep yourself going. This advice is true at any age.

If you're looking longer than you anticipated, figure out a way of using the time by doing something that you can put on your resume, whether it is taking courses, getting a certification, or doing an internship. You want to be able to account for the time in a way that expands your credentials and says you were able to keep your technical knowledge fresh.

Continue to plug along, even though you may find yourself repeating the same steps and wondering if they will ever pan out. They will. If you need fresh ideas for your search, check out The University of North Carolina at Wilmington, http://www.uncwil.edu/stuaff/career/Majors/.   "What Can I Do With a Major In?" a particularly good site to jog your thinking.

If you find yourself unmotivated, unwilling to spend the money to attend a professional conference or event, or even to go to a job fair, get out of bed and call a buddy to go with you. It is simply imperative to turn off the TV, get yourself off the couch, and keep on building your network.   In fact, just by staying in touch with a fellow graduate, going to that job fair, or picking up the phone, you might just run into someone who has a lead for you -- a little bit of information that will revitalize your search and get you going again.

Finally, if you find yourself wasting half the day, organize your time and put yourself on a schedule. Either get into a routine of waking up early and going to the gym, or decide to volunteer. You can always contribute your time at a public school helping someone learn to read, which will give you a new lease on life and positive re-enforcement regarding the kind of valuable contribution you can make to an organization.

Keep expanding your external network,
volunteer and do work at a school
and keep yourself busy and occupied
with your skills up to date.
Jay Colan, Senior Consultant/Vice-President,
Lee Hecht Harrison
New York, NY

Overall, don't be afraid to go out and keep trying. The skills that got you through school are also the very same skills that will help you find work in this tough job market.

APPENDIX:
ADDITIONAL SOURCES FOR YOUR JOB SEARCH

General Sites and Associations		
Web Site	**Site url**	**Our 2 cents**
Association of Computer Machinery (ACM)	www.acm.org	International computer association with academic computing and some other jobs. Members can also access 100+ free distance learning courses.
CareerBuilder	http://www.careerbuilder.com/JobSeeker/Jobs/jobfindit.asp	Considered a major career site of value for ALL job seekers, the IT section is notable for its International (Asia, Canada, Europe, etc) offerings. Search by company, field of interest, job title. Limited mainly to job searching/position posting, although there is resume creation help available.
CareerJournal	http://www.careerjournal.com	Related to the Wall Street Journal but not the same jobs. Considerable amount of career info and help; no single area devoted to IT.
Flipdog	http://www.flipdog.com	Excellent site to find many jobs and set up agents. Listings include Computer/MIS as separate category.

Hotjobs	http://www.hotjobs.com/	Good jobs for everyone - with separate high tech job categories - including (Web, Media, Technology). Good large and small listings . . . and now part of YAHOO! Excellent site for IT and business job postings.
Information Technology Association Of America (ITAA)	http://www.itaa.org	This is the place to go for the latest and greatest on News & Issues in IT. A definite must for anyone in the field. Resources include government and legal impact on the industry and salary data. Their "Bouncing Back" report is a must read for anyone in the industry.
IEEE	www.ieee.org	Excellent professional site that also links to www.computer.org, the site of the Computer Society. Dues in the Computer Society or in IEEE AND the Computer Society enable members to take distance-learning courses for free.
JOBWEB	http://www.jobweb.com/home.cfm	Career Site for National Association of Colleges & Employers (NACE); provides general info regarding Salaries, Hot Jobs, Employer Online Job Fair (contains company and contact info) and more.

		Archive of prior articles. This is a great place for undergrads and seniors to learn about a number of fields, including salary potential.
Monster	http://www.monster.com	The #1 rated career site overall. Though not geared specifically to IT, "3 of the top 10" job posting categories for Monster in 4-2002 were IT, engineering and computer software.
The National Center for Social Entrepreneurs	http://www.socialentrepreneurs.org	The National Center for Social Entrepreneurs
National Consortium of Entrepreneurship Centers (NCEC)	http://www.nationalconsortium.org	The National Consortium of Entrepreneurship Centers (NCEC), a non-profit organization representing more than 60 entrepreneurship centers at leading business schools in the U.S.
Open Door	www.keys2it.org	This non-profit formed by the National Association of Computer Consulting Businesses helps youth and adults of any age enter the IT profession.
Opportunity Knocks	http://www.opportunitynocs.org/index.jsp	Are you interested in non-profit work?

6FigureJobs.com	http://www.sixfigurejobs.com	For executive talent in ALL careers, but some of the IT groups profiled are; Cisco, Computer Sciences Corp (CSC), Unisys with direct email contact link.
SciTechResources.gov	http://www.scitech.gov	The site is in development but has over 1300 web sites listed providing the scientist, engineer, and "science aware" citizen with easy access to key government web sites. Communication Technology & Computer Science/Technology are the two IT major headings.
AWSEM	www.awsem.org	Started in 1994 as a project of Portland, Oregon's "Saturday Academy, AWSEM stands for Advocates for Women in Science, Engineering and Math and serves to help girls continue their studies in these fields.

BIOTECH/MEDICAL		
American Medical Informatics Association	http://www.amia.org/jobexch/fjobexch.html	The American Medical Informatics Association (AMIA) page for those who need a job or have a position to post.
Bioplanet	http://www.bioplanet.com/index.php	Bioplanet - a Job Searching/Recruiting resource for the rapidly growing field of Bioinformatics. Also contains basic and advanced articles in the field.
Bio.com	http://www.bio.com/	An excellent industry center for work in the biotech field, including IT.
The Shay Group	http://www.shaygroup.com/frame_index2.html	Medically and managed care oriented, includes permanent and contract positions with pharmaceutical companies, managed-care and insurance groups, medical device manufacturers, clinical research organizations and healthcare information software vendors.

EDUCATION		
Association for Information Systems	http://aisnet.org/placement	If you're interested in an IT position within education, check out Association for Information Systems placement page; fee-based for both job seekers and institutions.
Craigslist	http://www.craigslist.org	Not just for San Francisco, but check out craigslist for Atlanta, Austin. Boston, Chicago, Denver, DC, Los Angeles, New York City, Portland, Oregon, Sacramento, San Diego, Seattle, Vancouver, Sydney, and Melbourne. Updated DAILY!

GOVERNMENT		
Federal Jobs Central	http://www.fedjobs.com	For a fee you can see 100% of Federal Jobs
Federal Jobs Net	http://www.federaljobs.net	Lists 150 separate Federal agencies and links leading to their job listings.
National Security Agency (NSA)	http://www.nsa.gov/programs/employ/index.html	This is the official web site of the National Security Agency (NSA), containing an outline of the types of IT related/Computer related skills the department is looking for.
National Technology Transfer Center	http://www.nttc.edu/aboutnttc/jobopps.asp	Listing of IT and other positions available with the NTTC.
USAJOBS	http://www.usajobs.opm.gov/a6.htm	Lists about 60% of Federal jobs via the Office of Personnel Management.
US Department of Justice	http://www.usdoj.gov/06employment/06_5.html	Plenty of "Department of Justice" Government agencies are seeking high tech help, from the Federal Bureau of Investigation to the US Attorneys' offices to the Bureau of Prisons and the Immigration and Naturalization Service which protects our borders.

CERTIFICATIONS		
Brainbench	http://brainbench.com/xml/bb/homepage.xml	Lets you test online for any of 375 business, IT and management skills. Aside from general business and other skills such as language proficiency, writing, etc, there are two sections devoted entirely to Computer and IT related required functions. Otherwise, go to Oracle, Cisco, Microsoft directly for their particular education and certification sites.
Microsoft Certified Professional Magazine	http://www.mcpmag.com	Microsoft's News Magazine offers CERTIFICATION in using and deploying MS products.
TECHIES.COM - pure tech talent	http://www.techies.com	Very busy site, info is confusing!

JOB FAIRS, CONFERENCES, and CALENDARS		
BrassRing	http://www.brassring.com/	For high tech recruiting events, particularly in Silicon Valley
Entrepreneurial Events Calendar	http://www.entreworld.com/Calendar	Entrepreneurship events
The European IT Forum 2002	http://www.idc.com/getdoc.jhtml?containerId=IDC_P5094&selEventType=CONFERENCE	European IT Conference - **9/16-9/17**-Monaco; IDC's Flagship Event!
GARTNER'S Worldwide Interactive Calendar of Events & Conferences	http://www3.gartner.com/Events?pageName=calendar	Worldwide Interactive Calendar of Events & Conferences; many are available on the web..
HRLive	http://www.hrlive.com/	Sponsored by J Walter Thompson recruitment advertising, a good site for layoffs AND a job fair calendar
INT Media Events	http://www.intmediaevents.com/	IT & related fields events calendar - plus prior event tapes and conference proceedings.
Tech Web's Tech Calendar	http://www.techweb.com/calendar	IT shows worldwide
WORKIT	http://www.workit.com	An event calendar for SIGS and other technology events divided by region of the country. Recently started a separate section for women.

LAYOFF DATA		
The Layoff Lounge™	http://www.layofflounge.com	Go to a pink slip party
YAHOO NEWS-BUSINESS	http://dailynews.yahoo.com/fc/Business/Dot_Com_Shakeout; http://news.yahoo.com/fc?tmpl=fc&cid=34&in=business&cat=downsizing_and_layoffs	Yahoo's daily update on failed companies and re-priced options.

SALARIES		
COMPUTERWORLD Salary/Skills Survey	http://www.computerworld.com/departments/surveys/skills?from=bsm	Computerworld's 2000 Salary & Skills Survey; includes Job Title, Location, Industry, Company Size in $$. 2002 Survey data collection: June 3-29, 2002 Special report publishes on Oct. 28, 2002
JOBSTAR'S Computer & Engineering Salary Surveys	http://jobstar.org/tools/salary/sal-comp.cfm	Recent Computer & Engineering Salary Surveys & interactive salary calculator.
Network World Fusion's Salary Calculator	http://www.nwfusion.com/you2001/calculator.html	A wide range of high tech jobs from Vice President of IT to Telecommunications, Network Architect and General Management.

SALARY.COM	http://www.salary.com	Search by Job Category and Location via site's **Salary Wizard**. Includes detailed Job description. Salaries broken down by; Base, Health Care costs, 401k/403b, Bonuses, Disability Insurance, Social Security and Pensions to get total package value. Good data and more if you will pay for it.
COMPUTERJOBS. COM - Salary Ticker	http://www.ticker.computerjobs.com	Great listing of salaries in a number of IT fields; in addition you can Post Resume and Search for Openings.